Real Science-4-Kids

Laboratory Workbook

Pre-Level I

R.W. Keller, Ph.D.

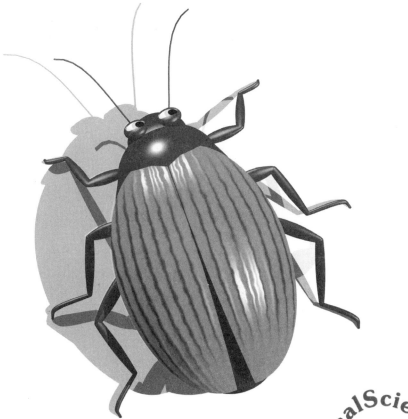

RealScience
4
Kids

Cover design: David Keller
Opening page: David Keller, Rebecca Keller
Illustrations: Rebecca Keller

Real Science-4-Kids: Pre-Level I Biology Laboratory Workbook

ISBN: 978-09799459-1-7

Published by Gravitas Publications, Inc.
P.O. Box 4790
Albuquerque, NM 87196-4790

Printed in United States

Gravitas
Publications Inc.
www.gravitaspublications.com

Special Thanks
to
Liam and Lillian
for their valuable input

A note from the author

Hi! In this curriculum you are going to learn the first step of the scientific method:

Making good observations!

In Biology making good observations is very important. This workbook has different sections. There is a section called "Observe it" where you will make observations. There is a section called "Think about it" where you will answer questions. There is a section called "Test it" where you set up an experiment to observe. There is a section called "What did you discover?" where you will write down or draw what you observed in the experiment. And finally, there is a sections called "Why?" where you will learn about why you may have observed certain things.

These experiments will help you learn the first step of the scientific method and... they're lots of fun!

Enjoy,
Rebecca W. Keller, Ph.D.

Contents

Experiment 1

Where does it go?

I. Observe it.

Observe the objects you collected.

Name the object and then describe it using words or a picture.

_____

_____ _____ _____ _____

_____ _____ _____ _____

_____ _____ _____ _____

_____

_____ _____ _____ _____

_____ _____ _____ _____

_____ _____ _____ _____

II. Group it.

Look at all of the objects you described. Think about different groups you might use to sort them. You might use **small** or **round** or **white** or **fuzzy**.

Name five groups you will use to sort the objects. Put each object in ONE group.

_____ _____ _____ _____ _____

_____ _____ _____ _____ _____

_____ _____ _____ _____ _____

_____ _____ _____ _____ _____

III. Think about it.

Are there objects that fit in more than one group? If so resort as many objects as you can into new groups.

_____ _____ _____ _____ _____

_____ _____ _____ _____ _____

_____ _____ _____ _____ _____

_____ _____ _____ _____ _____

Can you do it again?

_____ _____ _____ _____ _____

_____ _____ _____ _____ _____

_____ _____ _____ _____ _____

_____ _____ _____ _____ _____

III. What did you discover?

1. What did you observe about the objects you collected?

_____

2. Was it easy to pick groups to sort the objects? Why or why not?

_____

_____

_____

3. Was it easy to decide which objects would go in each group? Why or why not?

_____

_____

_____

4. The objects in a group have the same feature (round, or small). List some features that were different between objects in the same group.

The _____ objects were all _____ but some were also _____.
The _____ objects were all _____ but some were also _____.
The _____ objects were all _____ but some were also _____.
The _____ objects were all _____ but some were also _____.
The _____ objects were all _____ but some were also _____.

IV. Why?

It can be hard to sort objects into groups. Some round objects may also be fuzzy, like a cotton ball. And some other round objects might be smooth like a rubber ball. Some smooth objects might also be large. And some smooth objects might also be small. How do you decide which object to put in which group?

This can be a hard problem, even for scientists. Living things have lots of different features and it can be hard to figure out which living things go in which groups. Do you sort all the green creatures in one group and all the brown creatures in another group? This would be one way to sort green grass and bears. But what about a tree? A tree is both green and brown. Does a tree go with the grass or with the bears?

Scientists sometimes discover a new living thing—a creature they have never seen before. The first thing a scientist does is make careful observations about the creature. Is it green or grey? Does it have smooth skin or scaly skin? Does it live in the water or does it live on land? Does it eat vegetables or does it eat other animals? Can you see it with your eyes or do you need a microscope?

All of these observations help scientists know which group to put a new creature. By putting it into a group, scientists can better understand what is the same and what is different about the new creature compared to other creatures.

V. Just for fun.

You are an explorer and you find a new creature on a new planet.

It has the following features.

It is green. It eats flies. It lives in trees. And it flies with wings.

If you had to put your new creature into a group, would you group it with frogs, monkeys or butterflies?

_____

Draw a picture of this new creature.

Experiment 2

What do you need?

I. Think about it.

Cells are like little cities. There are lots of jobs that need to be done by lots of different workers inside a city. Your mom and dad also do lots of different jobs at your house. List some jobs your mom or dad do.

Job _____

Job _____

Job _____

Job _____

Job _____

Pick one of the jobs you listed and think about all the tools or items you think your mom or dad need to do this job. List the items you think they need.

_____

_____   _____

_____   _____

_____   _____

_____   _____

II. Observe it.

Draw a picture showing your mom or dad doing their job with the items they need.

Job _____

Pick one of the items your mom or dad uses to do their job. Draw a picture of the item with detail.

Item _____

For the item you drew, answer the following questions;

What is the item? _____

Where did the item come from?

_____

How did the item get there?

_____

Who made the item?

_____

What is the item made of?

_____

Where does the material that makes the item come from?

_____

Draw a picture showing all of the people you think are needed to make and deliver the item your mom or dad uses for their job.

III. What did you discover?

1. How many different jobs do your mom or dad do in your house?

_____

2. How many items does your mom or dad need to do the job?

_____

3. How many people does it take to make the item? List as many as you can.

_____

4. If your mom or dad had to make their own item to do the job, how many more jobs would they need to do? List a few.

_____

5. If your mom or dad had to make all of the items in your house how many jobs do you think they would have? List a few.

_____

6. Do you think it helps to have a city that can make certain items so your mom and dad can use the items to do other jobs?_____

IV. Why?

When you think about all of the items your mom and dad need to use to do certain jobs in your house, you find that they need lots of different items because there are lots of different jobs. It takes lots of different people doing lots of different jobs in different parts of the city or country to make the items your mom or dad use.

A cell works much in the same way a city works. There are places where different jobs get done and there are lots of different types of molecules that are needed for a cell to do those jobs. In order for a cell to live, a cell must make many of these molecules itself. In a cell, there are many molecules each doing different jobs to make all of the molecules a cell needs.

Each part of a cell has a different job to do. And each job a cell does takes different molecules. A cell must make sure all of the molecules it needs for living are in the right places at the right time and in the right amount.

When a cell does not have all the molecules it needs to do all the jobs it has to do, or when a cell does not have enough molecules, or when the jobs are not done in the right way, the cell cannot live. Just like a city would not work if people were not doing the right jobs, or were not in the right place at the right time, or if the jobs were not done in the right way.

Experiment 3

Who needs light?

I. Think about it.

List three things a plant needs to live?

_____

_____

_____

What do you think would happen to a plant if it did not get sunlight?

_____

_____

_____

_____

II. Observe it.

Take two small plants that are the same kind.

Carefully observe each plant.

List some words that describe the plants. Observe any unique features.

_____   _____

_____   _____

_____   _____

_____   _____

_____   _____

II. Observe it.

Draw each plant.

A

B

III. Test it.

Take the plant labeled A and put it into a sunny place.

Take the plant labeled B and put it inot a dark place.

Describe what you think will happen to each plant.

A _____

_____

B _____

_____

Note: Make sure you water each plant with the same amount of water.

Draw each plant after week 1

A

B

Draw each plant after week ———

A

B

Draw each plant after week _____

| A | B |

Draw each plant after week _____

| A | B |

IV. What did you discover?

1. How did the plants look the first day?

   Plant A _____

   Plant B _____

2. How did the plants look after the first week?

   Plant A _____

   Plant B _____

3. How did the plants look after the last week?

   Plant A _____

   Plant B _____

4. Describe any differences you observed between the two plants.

   Plant A _____

   Plant B _____

V. Why?

A regular houseplant needs sunlight to make food. If a houseplant is not able to get sunlight it cannot make the food it needs to stay healthy. Eventually a houseplant will die if it does not get enough sunlight.

When you put one plant in the dark and keep one plant in the sunlight, you are testing what happens to a plant that does not get sunlight. Why do you think you needed two plants? One for the sunlight and one for the dark?

You used two plants because as a scientist you want to make careful observations when you change something. When you use two plants (one in the light and one in the dark) you can easily compare the two plants. You want to know what happens to a plant in the dark and compare that to a similar plant that stays in the light.

This is called using a **control**. A **control** tells you what will happen if there are no changes. In this way a scientist can be sure that he can observe what happens when something is changed. In this case, you were observing what happened when sunlight was taken away from a plant. Your control plant (Plant A) told you what the plant would have looked like if it had sun. The plant you took away the sunlight from (Plant B) told you what would happen to a plant if it could not use the sun's energy to make food. Using a **control** helps you determine what happens when the sun is removed from a plant.

Experiment 4

Thirsty flowers?

I. Think about it.

If you put a white carnation into a glass of colored water, what do you think will happen to the flower?

_____

_____

_____

Draw a picture showing what you think will happen.

II. Observe it.

Carefully observe the carnation. Draw your observations.

Take one carnation and split it in half lenthwise. Draw your observations.

III. Test it.

Place the carnation in the colored water. Observe what happens to the flower. Draw what you see.

Start

After _____ minutes.

After _____ minutes.

After _____ minutes.

After _____ minutes.

Cut the stem open. Draw what you see.

III. What did you discover?

1. What did the carnation look like before you added the colored water?

_____

_____

2. What did the carnation look like inside?

_____

_____

3. What happened to the carnation when you put it into the colored water?

_____

_____

4. What did you observe in the stem after it was in the colored water?

_____

_____

IV. Why?

A carnation is a flower that usually has a long green stem. The stem is the part of the plant that brings in water and nutrients from the soil. The stem has tissues inside that act as little straws to draw water up from the soil.

When you place a carnation in colored water, you can observe what happens to the flower. Because the water is colored, you can watch the flowers "drink" the water from the glass. The colored water travels up to the top of the carnation and when it reaches the flower it starts to color the petals.

When you cut open the carnation, you can observe the tissues that move the water up the through the stem. When you cut open the stem, you can see if the tissues in the stem are colored too.

The tissues in a carnation are designed to move the water one way - up the stem. The water does not come back down through the stem. (Do you think that is true? Why don't you try it—take the stem out of the colored water and see if the colored liquid drops back out.)

A plant "drinks" water much like you drink water from a straw. Can you explain how a plant "drinks" water?

Experiment 5

Growing seeds

I. Think about it.

If you put a bean in a jar and add some water and let it sit for several days, what do you think will happen?

_____

_____

_____

Draw a picture showing what you think will happen.

II. Observe it.

Carefully observe a bean. Look at the outside. Draw your observations.

Take one bean and split it in half lenthwise. Draw your observations.

III. Test it.

Place a bean between a piece of white paper and the inside of a clear jar. Add some water and. Draw the bean and the jar.

Day 1

Day _____

Day_____

Day_____

Day _____

Day _____

III. What did you discover?

1. What did the bean look like when you looked inside? What did you find inside?

_____

_____

2. How many days did it take for the bean to start growing?

_____

_____

3. What started to grow first? Which way did it grow - up or down?

_____

_____

4. How many days did it take for the bean to turn into a seedling? In your own words describe briefly how it grew from a bean to a seedling.

_____

_____

IV. Why?

A bean is a seed. Seeds are how most plants begin. Inside of a bean you can see the embryo that will grow into the little plant, or seedling. Inside of the bean you can also see food the embryo uses to grows until it has roots and leaves to make its own food.

When you put a bean inside of the ground, it will start to sprout. You can watch a bean spout by putting it into a clear jar with water. The bean starts to sprout a root first. The root will grow downward finding its way to the ground. A root knows which direction to grow and it will not grow upward to the sun, but down into the ground. The shoot of the plant will grow next. It will grow upwards towards the sun so that when the leaves come out they can collect the sunlight.

The bean continues to grow the roots and shoot until it becomes a seeding. When it has leaves and a root big enough to gather nutrients is no longer needs the food it had inside the seed. It is a ready to become a big plant!

Experiment 6

Little creatures move

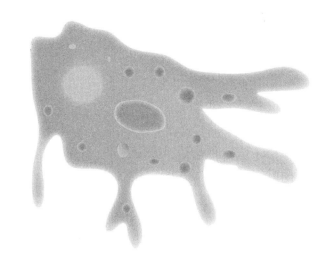

I. Think about it.

If you look at some pond water with a microscope, what do you think you will see? Draw what you think you will see.

II. Observe it.

Take some pond water and put it under the microscope. Draw what you see.

See if you can observe different moving creatures. Draw one here.

Draw a different moving creature here.

Draw a different moving creature here. Note if it moves differently.

Draw a different moving creature here. Note if it moves differently.

Are there two creatures that are similar? Draw them here.

Are there two creatures that are different? Draw them here.

III. What did you discover?

1. Did the pond water look like you thought it would? Why or why not?

_____

_____

2. What was the first thing you noticed about the pond water?

_____

_____

3. Was there anything you did not expect to find in the pond water? Describe it.

_____

_____

4. How many different creatures did you find? _____

5. Did you find two or more that moved in different ways? _____

6. Describe your favorite creature. Explain why it is your favorite?

_____

_____

IV. Why?

Pond water is full of little creatures. In fact, little creatures are found in soil, in hay, and in oceans and rivers. You might also find unwanted creatures on your toothbrush! Many of these little creatures are called protozoa.

There are many different kinds of protozoa. Many types of protozoa use a large whip-like tail called a **flagellum** to move. Other types of protozoa use small hairs called **cilia** to move. And other types of protozoa crawl using "false feet."

In a microscope you can observe protozoa move. You might see them move forward and backwards. You might see them bump into a piece of food or even bump into each other. You can see them roll, and stop, and turn, and then start moving again.

Protozoa need to move to find food, or escape from danger, or find a place to rest, just like you do. Humans have legs to move. If your mom calls you for dinner, you need to get up from reading this book and walk to the table to eat. If you are strolling in the park and big dog starts barking you might want to run away as fast as your legs will move. When it gets dark and you are ready for bed, you walk to your room (after using your new protozoa-free toothbrush on your teeth) to go to sleep. You are designed differently than a protozoa, but protozoa can also use their bodies to move just like you!

Experiment 7

Little creatures eat

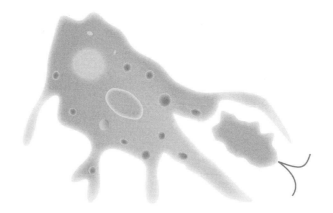

I. Think about it.

If you look at a protozoa eat with a microscope, what do you think you will see? Draw what you think you will see.

II. Observe it.

Take some pond water and put it under the microscope. Observe if any protozoa are eating. Draw what you see.

See if you can observe different eating creatures. Draw one here.

Draw a different eating creature here.

Draw a different eating creature here.

Draw different food one creature might be eating.

Are there two creatures that are eating in the same way? Draw them here.

Are there two creatures that are eating in different ways? Draw them here.

III. What did you discover?

1. Did the protozoa eat like you thought they would? Why or why not?

_____

_____

2. What was the first thing you noticed about the eating protozoa?

_____

_____

3. Was there anything you did not expect to find watching the protozo eat? Describe it. _____

_____

4. How many different ways did they eat?_____

5. Describe your favorite creature. Explain why it is your favorite?

_____

_____

IV. Why?

Protozoa eat in different ways. Some protozoa make their own food - like the green Euglena. Some protozoa use their tiny hairs to sweep the food into their mouths like paramecium. And other protozoa capture their food with their feet like the amoeba. Because there are lots of different kinds of protozoa there are lots of different ways protozoa eat.

Protozoa can eat lots of different kinds of food. They can eat algae or other small plants. They can eat yeast and they can eat other protozoa. Imagine what might happen when two protozoa-eating protozoa meet. Who gets to eat whom?

You also eat, but like most humans, you usually use your mouth to eat. Humans can eat both plants and animals. You usually don't need to hunt for your food - unless your brother steals your piece of cake! But you do need food. You are not like a Euglena who can make its own food and you can't catch your food with your feet like an ameoba. In a city, you need to rely on other people who can provide you with the food you need. You need milk from the dairy and bread from the bakery and eggs from the farm and just for fun—chocolate from the chocolate factory! A protozoa doesn't have other protozoa finding food for it—it must find its own food.

Experiment 8

Butterflies flutterby

I. Think about it.

How did the butterfly get its name? Is it because butterflies eat butter? Or do you think it is that butterflies are sometimes yellow—like butter?

Write or draw how you think the butterfly got its name.

How does a caterpillar turn into a butterfly? Draw or write how you think a caterpillar becomes a butterfly.

II. Observe it.

The beginning: The egg.

Draw and/or write what you see.

_____
_____
_____
_____
_____

The middle: The caterpillar

Draw and/or write what you see.

_____
_____
_____
_____
_____

The change: The chrysalis

Draw and/or write what you see.

The end: The butterfly.

Draw and/or write what you see.

Draw the life cycle of the butterfly as you observed it.

Was there anything you thought was very interesting? Draw it below.

III. What did you discover?

1. Did the butterfly eggs look like you thought they would? Why or why not.

_____

_____

2. Did the caterpillar look like what you thought it would? Why or why not.

_____

_____

3. Were you able to observe exactly what was happening in the chrysalis? Why or why not.

_____

_____

4. Did the butterfly look like what you thought it would? Why or why not.

_____

_____

5. Describe your favorite part of the butterfly life cycle. Explain why it is your favorite part.

_____

_____

IV. Why?

The butterfly is a creature that starts as something completely different—a caterpillar. If you never watched a butterfly egg turn into a caterpillar and a caterpillar turn into a butterfly you would not know that they are the same creature. It takes a keen eye and careful observation to find out how life goes for the butterfly.

How did the butterfly get its name? You probably found out that no one is quite sure. There are ideas about how the butterfly got its name but not everyone agrees. One idea is that the word butterfly comes from a very old word **buturfliog** which is a word of "butter" and "fly." But why butter? One idea is that butterflies like to eat butter and land on creamy, buttery foods in kitchens. The German word for butterfly means "milk thief" so maybe butterflies like butter and milk. But no one is really sure where the name for butterfly came from.

Sometimes disagreements happen even in science. We know much about the life of a butterfly because we can observe it. But scientists don't know everything because scientists can't observe everything. You were not able to observe exactly what was going on inside the chrysalis as the butterfly changed because you couldn't see inside the chrysalis. But you were able to make some observation. The most important job of a scientist is to make careful observations—and record exactly what they see - even if they can't see everything—just like you did!

Experiment 9

Tadpoles and frogs

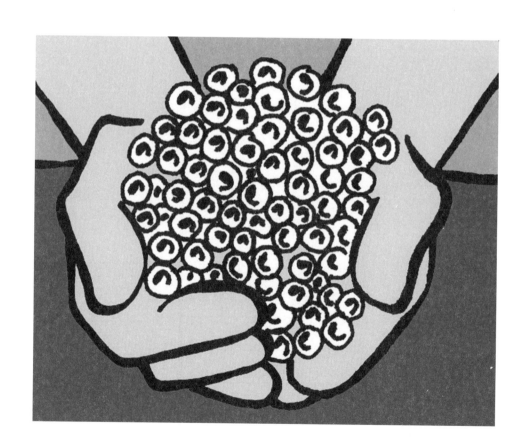

I. Think about it.

How did the frog get its name? Is it because frogs sit on logs? Or do you think it is that frogs have deep voices?

Write or draw how you think the frog got its name.

What do you think will happen as a tadpole turns into a frog? Write or draw what you think.

II. Observe it.

The beginning: The egg.

Draw and/or write what you see.

The middle: The tadpole eating.

Draw and/or write what you see.

The change: The tadpole with hind legs

Draw and/or write what you see.

_____

_____

_____

_____

_____

The change: The tadpole with front legs

Draw and/or write what you see.

_____

_____

_____

_____

_____

The end: The adult frog.

Draw and/or write what you see.

Draw the life cycle of the frog as you observed it.

III. What did you discover?

1. Did the frog eggs look like you thought they would? Why or why not.

_____

_____

2. Did the tadpole look like what you thought it would? Why or why not.

_____

_____

3. Were you able to observe exactly what was happening when the tadpole started to change ? Why or why not.

_____

_____

4. Did the adult frog look like what you thought it would? Why or why not.

_____

_____

5. Describe your favorite part of the frog life cycle. Explain why.

_____

_____

IV. Why?

A frog starts life as an egg and then becomes a tadpole before it becomes an adult frog. It changes significantly during its life cycle. A tadpole doesn't look like a frog, but more like a fish. But even though a tadpole may look like a fish, it is not a fish. If you never observed a tadpole changing into a frog you wouldn't know they are the same creature.

Some creatures like frogs and butterflies undergo a drastic change in appearance when they become adults. This process is called **metamorphosis**. Metamorphosis simply means "to change form or shape."

Humans do not undergo a metamorphosis as they grow into adults. Although you will look different when you are an adult than you do now, you do not completely change your form or your shape as you grow. You may get taller and your body will change proportions, but overall you keep your same shape and form you had when you were born. You are a different kind of creature than a frog or a fish!

Experiment 10

Creatures in balance

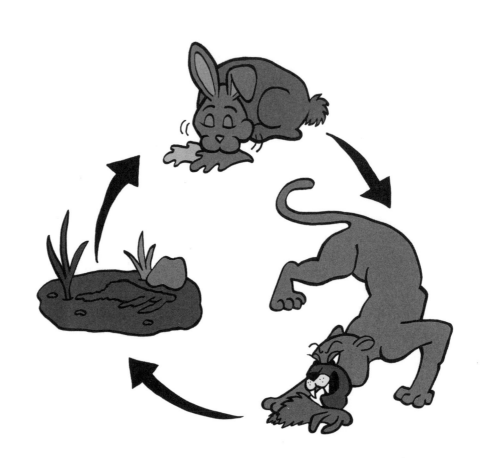

I. Think about it.

What would happen to the world if:
    there was too much water?
    there was too much sun?
    there were too few plants?
    there was too little oxygen?

Draw what you think the world might be like for one of these.

II. Observe it.

Day _____

Draw and/or write what you see.

Day _____

Draw and/or write what you see.

Day _____

Draw and/or write what you see.

Day _____

Draw and/or write what you see.

Day _____

Draw and/or write what you see.

Day _____

Draw and/or write what you see.

Day _____

Draw and/or write what you see.

III. What did you discover?

1. Describe what the ecosystem you built looked like on the first day?

_____

_____

2. Was it easy to keep the ecosystem healthy? Why or why not?

_____

_____

3. What problems did you have?

_____

_____

_____

4. What changes would you make?

_____

_____

IV. Why?

The earth is a delicately balanced ecosystem. It has different cycles that keep the it working in the right way. There is an air cycle, a water cycle and many food cycles.

Trying to create a small ecosystem can be very hard. It is a hard problem because it can be very difficult to find the right balance between the food, the air, the water, and the sun that are needed for life. And yet the earth is a system that maintains the balance needed for creatures of all sizes to live and grow, die and reproduce.

So far, no one has discovered another planet quite like ours. Our planet is unique in our solar system. Mars is too cold for the kind of life found on Earth and Mercury is too hot. Our planet is the only planet in our solar system that has just the right temerature for creatures to live and grow. Our planet also has just the right kind of atmosphere and just the right amount of water, and just the right amount of sun. Imagine how hard it would be to build an ecosystem like the earth!